Space Travel

by
Jenny Tesar

Heinemann
LIBRARY

First published in Great Britain by Heinemann Library
Halley Court, Jordan Hill, Oxford OX2 8EJ,
a division of Reed Educational & Professional Publishing Ltd

OXFORD   FLORENCE   PRAGUE   MADRID   ATHENS   MELBOURNE
AUCKLAND   KUALA LUMPUR   SINGAPORE   TOKYO   IBADAN
NAIROBI   KAMPALA   JOHANNESBURG   GABORONE
PORTSMOUTH NH (USA)   CHICAGO   MEXICO CITY   SAO PAULO

© Jenny Tesar 1997
The moral right of the proprietor has been asserted.

First published 1997

02 01 00 99 98
10 9 8 7 6 5 4 3 2 1

ISBN 0 431 01461 2

British Library Cataloguing in Publication Data

Tesar, Jenny
 Space travel. – (Space observer)
 1. Interplanetary voyages – Juvenile literature 2. Space
 flight – Juvenile literature
 I. Title
 387.8

This book is also available in hardback (ISBN 0 431 01460 4)
Printed and bound in Malaysia by Times Offset (M) Sdn. Bhd.

Acknowledgments
The publishers would like to thank the following for permission to reproduce
photographs:
Pages 4–5, 5 (inset), 6, 10-11, 13, 14, 15, 16-17, 20-21: Photri; pages 7, 12, 14, 18: ©NASA;
pages 8, 9, 19: Gazelle Technologies, Inc.; pages 22-23: A. Gragera, Latin Stock/Science
Photo Library/Photo Researchers, Inc.

Cover photograph: Photri

Every effort has been made to contact the copyright holders of any material reproduced
in this book. Any omissions will be rectified in subsequent printings if notice is given to
the publisher.

Contents

Some words are shown in bold, **like this**. You can find out what they mean by looking in the Glossary.

Going into space

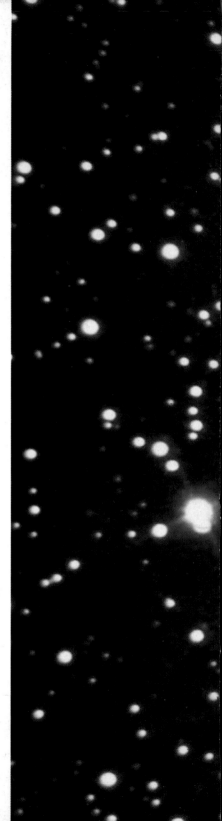

For thousands of years, people have wondered about space. They have dreamed of travelling to the stars and planets they could see in the sky.

The first person to travel into space was the Russian Yuri Gagarin. In 1961, he spent less than two hours in space. Since then, hundreds of people have gone into space. Some have travelled to the Moon.

Russian **astronaut**
Yuri Gagarin

Mission control

A trip into space is called a mission. People on the ground control the mission. They use **radio signals** and computers to stay in touch with the spaceship.

This mission control is at the Kennedy Space Centre

Astronauts are trained at mission control.

People at mission control also train **astronauts**. The astronauts learn how to wear space suits, move about and work in space, and operate their spaceship.

Leaving Earth

Spaceships have to be very powerful to escape Earth's **atmosphere**. They get their energy from big, powerful rockets. A rocket is a kind of engine.

Rockets help the spaceship take off

A spaceship heads for space

Spaceships are attached to rockets so they can be **launched**. When the rocket fuel is lit, the rocket and spaceship shoot up into space.

Space Shuttle

A space **shuttle** is a small spaceship that travels back and forth between Earth and space.

Space shuttles carry **astronauts** and materials into space. The astronauts do tests to learn how space affects their bodies. They fix damaged **satellites.** They also study Earth and other objects in space.

Two astronauts work in space while attached to their shuttle.

Living in space

Space is very different from Earth. In space there is no air to breathe. There is no water or food. Spaceships that carry **astronauts** also must carry air, water, food and things that people need to live.

An astronaut in space talks with mission control

Space suits protect astronauts against heat and cold

Spaceships and space suits protect astronauts against the very hot and cold temperatures in space.

Working in Space

Inside their spaceship, **astronauts** wear light, comfortable clothes. Sometimes, astronauts must go outside their spaceship to work. Outside, they need to wear a special suit made of many layers of strong material.

Astronauts wear regular clothing when working inside.

Astronauts wear space suits when outside the spaceship

The suit contains air to breathe. The helmet has earphones and a microphone so the astronauts can talk to each other and to mission control.

Space stations

A space station is a kind of **satellite**. It is a place where **astronauts** can live in space. Spaceships bring astronauts and supplies, such as food and water, to a space station.

In 1996, an American called Shannon Lucid lived in the Russian space station *Mir* for more than six months. She travelled 120 million kilometres and circled Earth 3008 times!

Shannon Lucid works with another astronaut at the *Mir* space station

Moon landings

American **astronauts** Neil Armstrong and Buzz Aldrin were the first people to walk on the Moon. They landed their **spacecraft**, *Apollo 11*, on the Moon on 20 July 1969.

Buzz Aldrin was one of the first astronauts to land on the moon

Astronauts have collected many moon rocks

After *Apollo 11*, there were five other landings on the moon. On most of these missions, astronauts collected lots of rock samples.

Space probes

Space probes are **unmanned spacecrafts** that explore outer space. Probes carry cameras, **radar**, and other instruments. They send information back to Earth.

Space probes have taken photos that scientists can use to make maps of space. In 1997, a space probe landed on Mars. It may tell us if there is life there.

The space probe *Voyager 2* is near Neptune and one of its moons

Space Colonies

Someday, people may live in space **colonies** that have homes, schools, and farms. The colonies will be under big domes filled with air.

Maybe you will be one of these space people. Will you work in a factory on a giant space station? Take a vacation on the Moon? The story of space travel is only just beginning!

This imaginary space base could grow to be a space colony!

Glossary

astronauts – people who travel to space

atmosphere – mixture of gases around a planet

colonies – communities of people who have left their home country to settle in a new place

launched – sent into space

radar – equipment that uses radio waves to find solid objects

radio signals – a wave of energy used to send sounds

satellite – a spacecraft that travels in a path in space

shuttle – something that moves back and forth from one place to another

spacecraft – a vehicle that travels to space

unmanned – without people

Index